THE SCIENCE OF ANIMAL MOVEMENT

How Bugs Jump

BY EMMA HUDDLESTON

CONTENT CONSULTANT
DAVID HU, PHD
PROFESSOR
MECHANICAL ENGINEERING
GEORGIA TECH

Kids Core
An Imprint of Abdo Publishing
abdobooks.com

abdobooks.com

Published by Abdo Publishing, a division of ABDO, PO Box 398166, Minneapolis, Minnesota 55439. Copyright © 2021 by Abdo Consulting Group, Inc. International copyrights reserved in all countries. No part of this book may be reproduced in any form without written permission from the publisher. Kids Core™ is a trademark and logo of Abdo Publishing.

Printed in the United States of America, North Mankato, Minnesota
042020
092020

Cover Photos: iStockphoto, foreground, background
Interior Photos: Mark Moffett/Minden Pictures/National Geographic, 4–5; Shutterstock Images, 6; NHPA/Photoshot/Science Source, 7; Maciej Olszewski/Shutterstock Images, 8; James Gathany/Ken Gage/Public Health Image Library/Centers for Disease Control and Prevention, 10; Stephen Dalton/Minden/Newscom, 12–13; Simon Shim/Shutterstock Images, 15; Nuwat Phansuwan/Shutterstock Images, 16; John Serrao/Science Source, 17; Red Line Editorial, 18; Ian Redding/Shutterstock Images, 19; Eye of Science/Science Source, 20, 29 (top); Christian Ouellet/Shutterstock Images, 22–23, 28; Stephen Dalton/NHPA/Photoshot/Newscom, 25; Michael Durham/Minden Pictures/National Geographic, 26; iStockphoto, 29 (bottom)

Editor: Marie Pearson
Series Designer: Ryan Gale

Library of Congress Control Number: 2019954238

Publisher's Cataloging-in-Publication Data

Names: Huddleston, Emma, author.
Title: How bugs jump / by Emma Huddleston
Description: Minneapolis, Minnesota : Abdo Publishing, 2021 | Series: The science of animal movement | Includes online resources and index.
Identifiers: ISBN 9781532192937 (lib. bdg.) | ISBN 9781644944325 (pbk.) | ISBN 9781098210830 (ebook)
Subjects: LCSH: Children's questions and answers--Juvenile literature. | Insects--Behavior--Juvenile literature. | Science--Examinations, questions, etc--Juvenile literature. | Habits and behavior--Juvenile literature.
Classification: DDC 500--dc23

CONTENTS

CHAPTER 1
Muscle Power 4

CHAPTER 2
Springing and Flinging 12

CHAPTER 3
Perfect Timing 22

Movement Diagram 28
Glossary 30
Online Resources 31
Learn More 31
Index 32
About the Author 32

Some species of katydids blend in with leaves.

CHAPTER 1

Muscle Power

A katydid feeds on a leaf in a tree. Of its six green legs, the back two stand out. They are much longer than the others. They help the katydid jump.

A bird sees the katydid. The bird swoops toward the bug.

Jumping is just one of a katydid's many defenses against predators.

At the last moment, the katydid's leg muscles tighten. Its back knees bend. Then it leaps from the leaf. The katydid has escaped.

Froghoppers are some of the many skilled insect jumpers.

Creating Thrust

Insects jump for many reasons. They do it to escape **predators**. They also jump to travel distances quickly.

Powerful legs help some insects, such as grasshoppers, jump far.

Bugs need thrust to jump. Thrust is a **force** that causes forward movement. Thrust helps bugs move against gravity. Gravity is a force that constantly pulls objects to Earth's center.

Grasshoppers create thrust with their legs. Other insects create thrust in different ways.

Insects can jump in three main ways. One is muscle power. Katydids rely on muscles in their long legs. The longer the leg, the more time it can be pressed to the ground during the jump. This is similar to how a kangaroo's legs let it jump.

Legs for Singing and Jumping

Grasshoppers and crickets jump with their long legs. But legs aren't only for jumping. Both insects can also rub their wings against their legs to make sounds.

Fleas are very tiny, but they can jump huge distances for their size.

There is a second way that insects jump. Many don't have long, powerful legs. Fleas and other insects have short legs. They rely on special body parts to store energy. They release that energy quickly to leap. Finally, other bugs such as grasshoppers combine long legs and stored energy to jump.

Further Evidence

Look at the website below. Does it give any new evidence to support Chapter One?

Insects Pictures & Facts

abdocorelibrary.com/how-bugs-jump

Special body parts help some bugs, such as the leafhopper, jump far.

Springing and Flinging

Insects with short legs have special body parts. These parts work like springs. They bend or snap into certain positions. They release quickly to allow a bug to jump.

Energy is what powers springing movements. **Elastic** body parts such as knees can store up energy by either stretching or being pressed down. A notch or peg holds the spring in this position. The stored elastic energy is held for later use. When the spring is released, the stored energy flings the body into the air like a catapult.

Fleas Use Protein

Fleas push off the ground through their shins and feet. In one of their muscles is a protein called resilin. Resilin holds energy. When fleas push off the ground, the movement releases energy in resilin, causing them to spring upward and forward. Fleas' shins and feet are covered in stiff, hairlike structures. The hairs help fleas grip the ground before jumping.

The narrow spot in the middle of the click beetle's body helps it jump when lying on its back.

Short-Legged Jumpers

One jumping bug is the click beetle. Click beetles roll on their backs when scared. They play dead. When they are ready to get back on their feet, they jump. They don't need their legs to jump. They use their bodies.

A click beetle doesn't always land on its feet when jumping from its back. Sometimes it takes a few jumps.

Click beetles have three body **segments**. They have a head, a thorax in the middle, and an abdomen at the rear. There is a **hinge** between the thorax and abdomen. To jump

Click beetles play dead so other animals don't try to eat them.

from its back, a click beetle bends its head and thorax off the ground and toward the abdomen. A long muscle under the shell connects the thorax and abdomen. When stretched, the muscle tightens and stores elastic energy. The hinge is secured with a peg.

Click Beetle Hinge

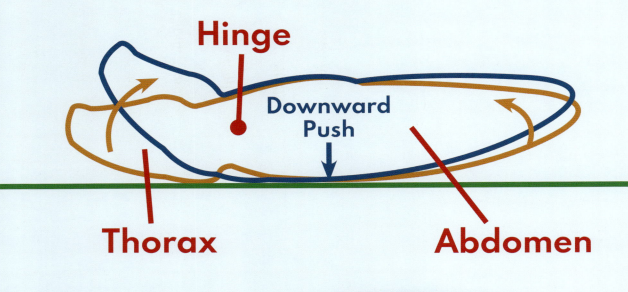

A click beetle stretches the muscle under its shell straight, storing energy. When released, the beetle's back snaps into a V position. This launches the beetle into the air.

When the beetle is ready to jump, the peg slides open. It unlocks the hinge. This makes a clicking sound. The muscle quickly snaps back to its original length. The head and abdomen move toward each other in a V. The beetle's

Springtails are a kind of very small bug that does not have wings.

back pushes against the ground, launching the beetle. It flips through the air and lands on its feet.

Springtails are tiny jumping bugs. They are only a fraction of an inch long. They don't have wings. Springtails get their name from how they jump. A springtail has a tiny body part called a furcula on its rear. The furcula is like a tail. Most of the time, it is tucked up under the body. A clasp keeps it in place. It stores energy.

A magnified image shows a springtail's furcula extended after a jump.

When the furcula is let go, it strikes the ground. The forceful hit flings the springtail into the air. The bug can jump several inches into the air. Springtails jump to escape predators.

Primary Source

Gregory Sutton studies fleas in the United Kingdom. He said:

> If you see fleas . . . jump and you realize how small they are, it doesn't take much to realize these guys are catapulting themselves huge numbers of body lengths.

Source: Wynne Parry. "For High-Jumping Fleas, the Secret's in the Toes." *Live Science*, 10 Feb. 2011, livescience.com. Accessed 12 Feb. 2020.

Comparing Texts

Think about the quote. Does it support the information in this chapter? Or does it give a different perspective? Explain how in two to three sentences.

A grasshopper's rear legs have more muscle than its other legs.

Perfect Timing

Some insects, such as grasshoppers, use a combination of muscle power and stored energy to jump. They need perfect timing to jump successfully. Otherwise, they might not jump as high as they need.

Grasshoppers have long legs. They have strong muscles, but they also use stored energy to power their jumps. First, a grasshopper bends its back legs in toward its body. Its long leg muscles are stretched in this position. The position puts pressure on its knee **joint**. An elastic **tissue** in the knee called the cuticle stores energy.

Jumping Spiders

Jumping spiders are not bugs. But they are great jumpers too. The spider tightens muscles at the top of its legs near its body. With these muscles tightened, blood can flow into the legs, but not out. The blood pressure increases and forces the spider's legs to straighten. All the legs snap to full length at once. The spider launches into the air.

Grasshoppers push against a surface with their rear legs to jump.

To jump, the grasshopper quickly relaxes its muscles. The cuticle releases the energy. The grasshopper's legs straighten as it springs into the air. Its muscles provide 40 percent of the power behind its jump. The other 60 percent comes from stored energy in its cuticles.

Sometimes grasshoppers jump into the air to begin flying.

Crash Landings

Insects that jump with muscle movement can control where they go and how they land. Grasshoppers in part use stored energy to jump. Their jumps tend to be out of control. But grasshoppers have hard exoskeletons. These exoskeletons protect their bodies from getting hurt in crash landings.

Bugs have different methods that help them leap high and far. Whichever method they use, bugs are some of the strongest jumpers in the animal kingdom.

Explore Online

Visit the website below. What new information did you learn about insects that wasn't in Chapter Three?

Grasshoppers and Relatives

abdocorelibrary.com/how-bugs-jump

Movement Diagram

Grasshopper

- Uses a combination of long legs, strong muscles, and stored energy
- Relaxes muscles to release energy stored in muscles and knees, launching into the air

Springtail

- Uses short legs and stored energy
- Builds energy in the furcula, which releases and hits the ground

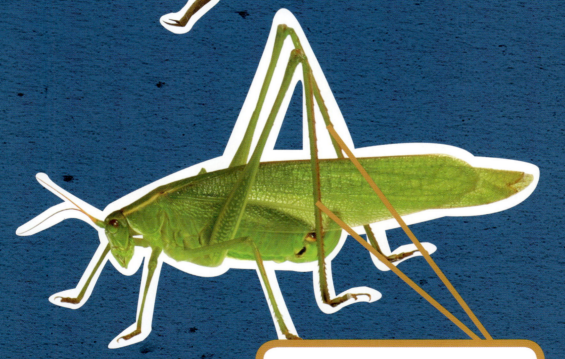

Katydid

- Uses long legs and strong muscles
- Pushes off ground to jump

Glossary

elastic
stretchy and able to snap back into place

force
an action that can start, change, or stop an object's motion

hinge
the meeting point of two stiff parts that allows them to swing back and forth

joint
a place where two parts of a skeleton meet

predators
animals that hunt other animals for food

segments
sections or parts

tissue
a group of one kind of cell that forms a structure in a living thing

Online Resources

To learn more about how bugs jump, visit our free resource websites below.

Visit **abdocorelibrary.com** or scan this QR code for free Common Core resources for teachers and students, including vetted activities, multimedia, and booklinks, for deeper subject comprehension.

Visit **abdocorelibrary.com** or scan this QR code for free additional online weblinks for further learning. These links are routinely monitored and updated to provide the most current information available.

Learn More

Hamilton, S. L. *Crickets.* Abdo Publishing, 2019.

Mills, Andrea. *Bugs.* DK, 2017.

Polinsky, Paige V. *Super Simple Experiments with Forces.* Abdo Publishing, 2017.

Index

click beetles, 15–18
crickets, 9
cuticles, 24–25

elastic body parts, 14, 17, 24
energy, 11, 14, 17, 19, 23–26

fleas, 11, 14, 21
furcula, 19–20

grasshoppers, 9, 11, 23–27
gravity, 8

joints, 24
jumping spiders, 24

katydids, 5–6, 9

muscles, 6, 9, 14, 17–18, 23–26

resilin, 14

springtails, 19–20
Sutton, Gregory, 21

About the Author

Emma Huddleston lives in the Twin Cities with her husband. She enjoys reading, writing, and swing dancing. She thinks the science of animal movement is fascinating!